To Elizabeth,

With love

Michelle
xx

Peaceable Kingdom

The Family of Animals
Peaceable Kingdom

By Ann Guilfoyle

Commentary by Edward R. Ricciuti

BOOK CLUB ASSOCIATES
LONDON

First published in 1980

This edition published 1980 by Book Club Associates by arrangement with George Allen & Unwin Ltd.

Printed in the United States of America

Color reproductions produced at Studio Analysis under the personal supervision of Paolo Riposio, Torino, Italy. Represented in the United States by Offset Separations Corporation, New York.

DEDICATION

To all families
everywhere
but especially my own.

CONTENTS

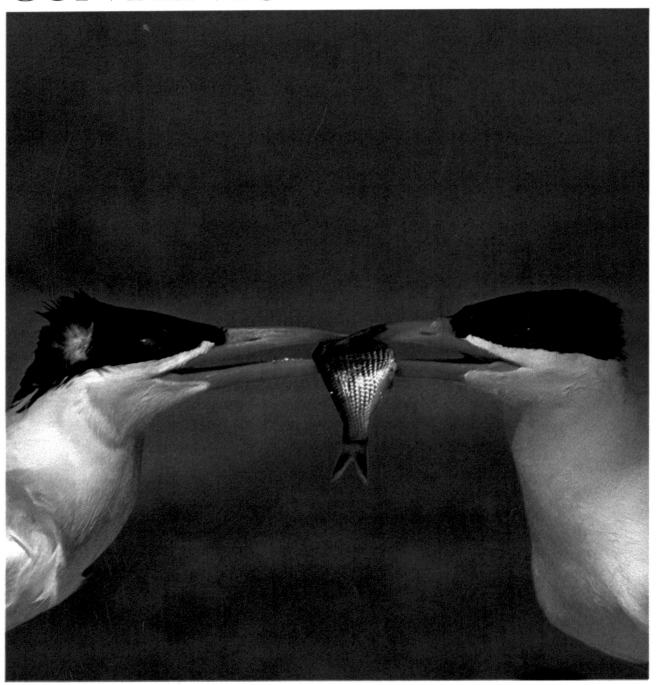

Acknowledgments 9

Introduction 10

1 Rivalry and Ritual 16

2 New Generations 36

3 Ties That Bind 56

4 Play and Perils 76

5 Checks and Balances 90

6 The Closed Circle 108

Afterword 131

Captions 133

ACKNOWLEDGMENTS

No picture book is created by a single person, but *The Peaceable Kingdom* more than most has been a cooperative venture.

I thank, first and always, the photographers whose work sustains my life and delights my soul.

I thank George Hornby, my father, who knows more about the making of a book than anyone. For my entire professional life, his knowledge has never been farther away than the other end of a telephone. He has taught me many things, not the least of which has been an honourable craft.

I thank Robert R. Reid, designer, who put his good eye to a difficult concept and helped to make it real. For reasons beyond anyone's control he did not finish, and I thank Don Werner who is responsible for the final, beautiful design of *The Peaceable Kingdom*. His well-felt blend of space, type and photograph is too rare in the world of picture books.

I thank Edward R. Ricciuti for his excellent text and his goodwill in helping a friend along.

I thank Jim Hoffman, master-of-words, for his expert copyediting.

I thank Toni Lopopolo of Macmillan for sharing a vision of *The Peaceable Kingdom* and giving the book its chance.

I thank Lois DeLaHaba, friend and literary agent, who treated this book as though it were her own and who made a hard time fun.

I thank Abner Sundell, friend, whose support was the difference between an idea and a finished book. Without his stubborn strength behind it, *The Peaceable Kingdom* would not have been.

INTRODUCTION

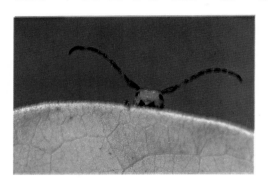

Time and seasons cycle for ever. Days and nights. Springs, summers, autumns, winters. Dry months and wet. Tides high and low. And so, too, the earth's creatures move through the intimate cycles of birth, mating, death. Populations expand and decline. There are periods of plenty and of famine. Rest alternates with activity. Cycles within cycles, independent yet interacting, the large encompassing the small, the small fusing into the large. Each moment of life is shaped by everything that has gone before. No act or circumstance stands alone; each has been influenced by the past and, in turn, will affect the future. There is no living thing, no biological or environmental event, condition or process, that deviates from this universal principle. Interdependence is the supreme law of nature.

It is through the actions and behavior of wild animals—from microscopic mote to massive leviathan—that these cycles of nature manifest themselves most clearly. Within a few decades, a patch of woodland in East Africa can be reduced to a savanna by elephants browsing and bulldozing their way through the acacias and other trees. With the trees vanished, long grasses take over, and the elephants move on to other woodlands that have grown in their absence. Or, at least, that is how it happened in the past. Today, confined within the unnatural boundaries of national parks, the tall beasts have nowhere else to go. The cycle has been broken by human intervention. After the elephants eat the trees and then the grass, there is no food left—and they starve.

Every ten years or so, the snowshoe hare population peaks. Then it declines, abruptly, in response to a diminished food supply. The Canada lynx—dependent on the hare for its sustenance—will undergo a similar cycle in population, a year behind its prey. The predator controls the prey; the prey controls the predator.

The actions and behavior of the members of the animal kingdom are interrelated with temporal and seasonal cycles. The round of day and night, for instance, is mirrored in the wakings and risings that are timed by the internal body clocks of animals. Day rouses the wasps, bees and butterflies. The light sends the hawks and eagles soaring and sets the songbirds singing. Hummingbirds hover about open blossoms. Army ants march. Chipmunks feed. Coatimundis troop over the ground and up into the trees in search of things to eat.

Night brings rest for the creatures of the day and a rush of activity to a new gathering of creatures. Raccoons prowl stream edges. Flying squirrels glide between trees. Leaf-cutting ants ascend tree trunks and clip foliage that will serve as compost for their underground fungus gardens. Owls noiselessly patrol their hunting grounds. And moths flutter on silent wings.

Creatures that thrive when it's light, creatures that flourish when it's dark, share the same habitat—an environment that supports two separate but interdependent systems of life, one adapted to day, the other to night. Both occupy and use the same physical space.

Changes of seasons are also directly reflected in the cyclical behavior of animals. Early in the year, rains turn the Serengeti Plain of East Africa into a lush, rich feeding ground where vast herds of zebras, wildebeests, hartebeests and other game swarm across the landscape. Here, with food plentiful, the young will be born, protected by their very concentrations of numbers from the predators that follow the herds. By July, the rains are but a memory, and the plain is dusty, barren and empty. The vast herds break down into smaller groups and drift to the north and west where wooded hills will still support forage and provide water.

Autumn in the North Temperate Zone sends many species of birds south and brings others down from the north. The orioles, redwing blackbirds and warblers depart. The juncos, evening grosbeaks and pine siskins arrive. Next comes the snow and ice of winter. By the time it ends, both groups of birds will be shifting northward once more. Frogs will be singing as the spring arrives. By summer, there will be young birds in the nests and tadpoles in the ponds and puddles.

To the human intelligence these cycles seem exact. Each season—and the processes and activities it sets in motion—appears to be neatly boxed as if on a calendar, distinct from the others. But in reality there are no neat separations. To an animal the only beginning or end is birth or death—seasons are a continuum. Like links in a chain, seasonal activities merge into one another. When to human senses a season seems to be at its peak, animals already are behaving in response to stimuli that will propel them into the activities of the next season. Thus, when summer is at its hottest, shorebirds are already moving south from the Arctic, fleeing from blizzards that are yet to come. In August, the monarch butterflies of the northeastern United States and southeastern Canada have grown restive and some have already begun their flight to Mexico. In winter, raccoons seek mates; their young will be born in the spring when the weather is mild and food plentiful.

It can be said that one animal is really many different animals, in behavior, physiology and physical appearance, not only at various stages of its life cycle but also at different times of the year. The bull elk that tosses its antlers and bugles its fierce autumn challenge will be uncrowned, furtive and docile within a few months.

A host of external cues—especially day length, temperature and rainfall—trigger complex and subtle changes that stimulate life-producing and life-sustaining activities such as breeding and migrating. Internal stimuli, also cyclical, set in motion the ways in which various animals struggle for territory, conduct rituals of courtship, mate and give birth, rear offspring, prepare for the separation of young from the parents, mature, age and die.

The manner in which natural cycles intermesh is complex beyond our understanding. We can glimpse only reflections of its intricacy. Clearly, however, the complexity of nature is not chaotic, but profoundly ordered, so that all parts work together to maintain an equilibrium. Thus balanced, undisturbed by humanity, nature is truly at peace.

When one cycle of existence draws to completion, the forces that bring it to a close generate new life, resuming the cycle. Life and death are one in the world of nature. In the death of a living being is the source of life to come, whether an animal dies to nourish another, or its carcass returns to the soil, where the myriad organisms of decomposition begin the work that will convert nonliving matter into the stuff of existence. It is not a case of a definite ending and a fresh beginning. Rather it is one of the new emerging from the old, with no dividing line between where one stops and the other starts, for this is the cycle of the universe.

Edward R. Ricciuti

The Peaceable Kingdom

For the animal shall not be measured by man.
In a world older and more complete than ours
they move finished and complete, gifted with
extensions of senses we have lost or never
attained, living by voices we will never hear.
They are not brethren, they are not underlings,
they are other nations, caught with ourselves
in the net of life and time, fellow prisoners of
the splendour and travail of the earth.

Henry Beston

THE PLATES

Yellow-billed Storks *(Mycteria ibis)*. Africa. M. Philip Kahl.

Kob Antelopes *(Kobus kob)*. Africa. Arthur Christiansen.

Gray Tree Frog *(Hyla versicolor)*. United States. Larry West.

Northern Elephant Seal *(Mirounga angustirostris)*. Baja California. The Sea Library—Robert T. Evans.

Pronghorns *(Antilocapra americana)*. United States. G.C. Kelley.

Broad-winged Damselflies *(Agrion splendens)*. England. Natural History Photo Agency—Stephen Dalton.

Northern Elephant Seals *(Mirounga angustirostris)*. Baja California. The Sea Library—Robert T. Evans.

Flap-necked Chameleons *(Chamaelo dilepsis)*. Africa. Anthony Bannister.

Skipper Butterflies *(Hesperiidae)*. United States. Zelda Glasser.

Roman Snails *(Helix pomatia)*. England. Natural History Photo Agency—K.P. Mafham.

Dolphin Gull *(Leucophaeus scoresbii)*. Argentina. Francisco Erize.

Lions *(Panthera leo)*. Africa. Grant Heilman.

For, lo, the winter is past,
The rain is over and gone,
The flowers appear on the earth,
The time of the singing of birds is come,
And the voice of the turtle is heard in our land.

The Song of Solomon

Now
is the hour
for the dauntless
spirit . . .
now for the
stout heart.

Virgil

The universe resounds with the joyful cry, I AM!

Scriabin

Oh the earth was made for lovers, for damsel, and hopeless swain,
For sighing, and gentle whispering, and unity made of twain,
All things do go a courting, in earth, or sea, or air....

Emily Dickinson

O, thou art fairer than the evening air
Clad in the beauty of a thousand stars.

Christopher Marlowe

For if she will, she will;
You may depend on't.
And if she won't, she won't;
So there's an end on't.

Anonymous

Whenever an infant is born, the dice . . . are being rolled again. Each one of us is a statistical impossibility around which hover a million other lives that were never destined to be born.

Loren Eiseley

Into the dangerous world I leapt,
Helpless, naked, piping loud.

William Blake

*For unto us
a child is born.*

Isaiah

3 Ties That Bind: *Care of Young*

In the animal world the duties of caring for the young are assumed in some species by just one parent and in other species by a widely extended family, with various kinds of familial relationships and arrangements ranging between these two extremes. The anthill and the beehive, for example, house super-families, in which the labor of nurturing the young is shared by multitudes. The female European earwig insect handles the task alone, and does admirably. She is kept busy licking her young clean and retrieving any strays that wander away from the miniscule den she has dug for her family, while with sharp pincers, she fiercely fends off intruders that might menace her progeny.

To most humans the king cobra of Southeast Asia, world's largest venomous snake, might seem the most unlikely creature to be a good parent. The female cobra, however, is a solicitous mother, at least until her eggs hatch. She builds a nest by using her coils to push and drag leaves together into a pile. After she lays her eggs, she covers them with more leaves and, while they are incubating, spends most of her time curled in the middle of the nest, always on her guard.

Care of the nest is parental stewardship at its most basic, but it still can be an exhausting job. The male megapode, an Australian groundbird, works almost year-round maintaining a mass of rotting vegetation in a pit a yard deep and more than a dozen feet in diameter. Only after all the eggs—sometimes dozens—hatch does he have a few weeks off before nest maintenance begins once again. The young are ready to care for themselves almost immediately after hatching, so they have no real contact with either parent.

However, among most of the birds—and the mammals—parental care reaches its highest stages of development. Not only do they feed and guard their offspring but also they instruct them in the lessons they need in order to survive. When young ospreys are ready to try their wings, the parents flutter and circle around their huge nest, screaming high-pitched calls that seem to encourage the fledglings to imitate adult flying. The mother raccoon leads her brood on nocturnal forays along watercourses and by lakesides. The young coons watch as their mother pulls crayfish out from under rocks and plucks mussels up from the silt, and soon they can do it too.

Many bird parents work so hard to feed their young that they themselves scarcely have time to eat. But for most birds, parental responsibility is rather short-lived. Robins tend their nestlings for only a few weeks. Red-shouldered hawks nurture their offspring for less than two months. Parenthood can also be a short-term affair among the smaller animals, although many mice, shrews and similar creatures may produce several litters yearly. With larger animals, the period during which the young depend on their parents lasts considerably longer. The female black rhinoceros stays close to her calf for about two years. A young gorilla still may demand some maternal attention more than four years after birth; during this prolonged rearing period, the gorilla youngster has plenty of time to learn the complex social behavior required of such an advanced primate.

However long the family group exists as an interdependent unit, it is held together by relationships that have come to be called "bonds." Many types of bonds exist in the animal kingdom—those between mates, for example—but in those species that maintain family life none are stronger than the ties that link parent and offspring. Familial cohesion is created in various ways. Waterfowl hatchlings undergo a process called "imprinting," in which they identify with the first large, moving shape they see. In the nest, this almost always is their mother. The bleating of a newly born lamb causes its mother to lick and nuzzle it, which in turn results in the infant sheep's feeling close to, and dependent upon, this constant source of warmth and comfort. Sound, scent, touch and sight interplay in an elaborate cyclical system that builds and reinforces bonds between parents and offspring throughout the period that the youngsters need in order to be able to survive.

THE PLATES

Greater Flamingoes *(Phoenicopterus ruber)*. Africa. Bruce Coleman, Inc.—M. Philip Kahl.

Vervet Monkeys *(Cercopithecus aethiops)*. Africa. Arthur Christiansen.

Lions *(Panthera leo)*. Africa. Edward S. Ross.

Cape Hunting Dogs *(Lycaon pictus)*. Africa. Anthro Photo—James Malcolm.

Green Herons *(Butorides virescens)*. North America. Leonard Zorn.

Shelducks *(Tadorna tadorna)*. Europe. Arthur Christiansen.

Polar Bears *(Thalarctos maritimus)*. North America. Fred Bruemmer.

Hippopotamus *(Hippopotamus amphibius)*. Africa. Bruce Coleman, Inc.—Norman Myers.

Kodiak Bears *(Ursus middendorffi)*. North America. Jeff Foott.

Orangutans *(Pongo pygmaeus)*. Malaysia. Anthro Photo—Richard Wrangham.

Vervet Monkeys *(Cercopithecus aethiops)*. Africa. Anthro Photo—Joseph Popp.

African Elephants *(Loxodonta africana)*. Africa. Animals, Animals—Al Szabo.

I feel
The link of nature draw me; flesh of flesh,
Bone of my bone thou art....

John Milton

nothing which we are to perceive in this world equals
the power of your intense fragility.

e. e. cummings

She is a tree of life to them....

Proverbs 3:18

May you grow old;
May your roads be fulfilled;
May you be blessed with life.

Pueblo Indian

To be young was very heaven....
William Wordsworth

To lift one if one totters down,
To strengthen whilst one stands.

Christina Georgina Rossetti

Child of the pure, unclouded brow
And dreaming eyes of wonder.

Lewis Carroll

And I saw in the turning so clearly a child's
Forgotten mornings....

Dylan Thomas

Born for success he seemed,
With grace to win, with heart to hold,
With shining gifts that took all eyes.

Ralph Waldo Emerson

Child you are like a flower,
So sweet and pure and fair.

Heinrich Heine

*For each age
is a dream
that is dying,
or one
that is coming
to birth.*

Arthur O'Shaughnessy

120

Is it so small a thing
To have enjoyed the sun,
To have lived light in the spring?

Matthew Arnold

*WHO CAN TELL
WHEN ONE SEASON ENDS
AND ANOTHER BEGINS?*

John Burroughs

RETROSPECTIVE

AFTERWORD

The Peaceable Kingdom was created at the time that I was experiencing the joyous sense of family that came to me when my first grandchild was born. It represents my celebration of new beginnings through new generations and I gladly share it with you. While this is a book about animals, it is about animals as they are perceived through human eyes, and the significances given to their behavior are human significances. It is, in truth, a book about human concerns and hopes. In many ways our needs are not very different from those of the animals pictured in these pages. It is to these human needs and hopes and concerns that I address my afterword.

All living creatures share certain basic characteristics that are the rules of life on this planet. The most basic of these manifest themselves as instincts and are forces that operate beneath consciousness to shape patterns of behavior. We humans tend to feel that we are free from instinct as a binding force. We delude ourselves. We hold in common with all other creatures at least one basic primeval drive: the simple, undeniable urge to survive, not just as an individual, *but as a species.* Most altruism among wild animals, most cooperation, most social behavior, in one way or another furthers mating and the rearing of young. Even when altruism of any kind is absent from a species' makeup and when social behavior does not exist at all, systems have evolved to guarantee that there will be new life to replace the parent. The thrust of life is to perpetuate itself through new life. It is as true for humans as for animals.

The life-force manifests itself in many complicated ways and we have—in all of our cultures and from the beginning of our time—acknowledged its existence and paid it tribute. But this is not enough. Our home—the earth—is a mess. Many of the animals that share it with us are endangered or are disappearing because of the pressures of human needs and carelessness. Every day, ignoring the potential consequences of modern warfare and ecological ruin, we diminish our own children's best chances for future survival. We go about our lives raising our families, suffering through the problems of everyday existence, hoping that things will take care of themselves. And while we turn the other way, these "things," which we would rather not confront, take hold and grow worse.

It becomes crucial that we re-examine the goals that we spend our lives striving to attain. If our planet is to renew itself, we must look at our personal priorities in terms of real need and be prepared to do with less. And the bombs that are not of our choosing, nor of our making, must be defused. Our leaders can no longer play king-of-the-mountain, making decisions that do nothing to further the progress of daily life. It is not too late to restore our good earth and clean skies and clear waters. And our children are entitled to sleep at night without fear of nuclear annihilation. In no way do I wish to imply a return to the garden. We live in a technological age and face a future that should be bright and good, incorporating technology to enhance life. But as one of the imperatives of all life is a viable environment that allows life to continue, our technology of the future must accommodate itself to the environment.

It is time that we looked around to recognize ourselves for what we are—a great family, a family of man—bonded together by common concerns that go beyond any secondary bonds of race, social class, country or continent. Beneath all the complexities of our individual existences there lies a simple and deep-seated desire to live well and to know that our offspring will have the opportunity to do the same when we are gone. We share this with all other residents of this earth, but we alone bear the burden of knowledge—and the responsibility that must accompany this knowledge.

It is time that we ordinary folk, already deeply committed to the concept of family, set out to guarantee a future for those we love. Armed with good sense and common purpose, fortified by the very vastness of our numbers, surely each one of us can save one small part of this, our home planet. I suggest to you that, once begun, this would be a movement the like of which has never been. I further suggest that, harnessed and functioning as a whole, the combined powers of the multitudes of "ordinary" men and women who make up the species *Homo sapiens* are a force for peace to cause angels to blush and the heavens to sing. I give you *The Peaceable Kingdom.* Once, a long time ago, we, too, were promised a place within it. It is past time that we claimed our birthright.

Ann Guilfoyle

CAPTIONS

1 Rivalry and Ritual

The males of many species of birds arrive at the breeding grounds before the females to establish claims to nesting territories. Such is the case with the yellow-billed stork. One male has taken possession of a nest site in a tree, but to keep it he must ward off another male that is about to land. Later, the females will appear and one will pair off with whichever of these rivals has managed to hold the mating place.

M. Philip Kahl

Two impala bucks square off for combat over a mating territory. Whichever buck wins this area of land will have the right to mate with members of female groups that wander into it. He will try to keep them within his territorial boundaries for as long as he can. Once the breeding season passes, however, the bucks lose their desire to be lords of the land and even join bachelor herds, indifferent to the females.

Arthur Christiansen

Larry West

A male gray tree frog, vocal sac inflated, sends his loud, rattling mating message into the night. During the breeding season, the males call from trees and shrubs on the margins of ponds and marshes and also from the edges of the water. Almost every kind of frog and toad utters a distinctive sound, which attracts females of the same species. The males are extremely vocal in the breeding season, but generally silent during the rest of the year.

Robert T. Evans

A bull northern elephant seal challenges a male sexual rival. Generally, the dominant bulls win such encounters merely by bellowing. But sometimes actual fighting occurs, especially when the contestants are two subordinate males, neither of which has claims to mastery. Although rarely fatal, the battles between these huge, aquatic, carnivorous mammals are very bloody, yet at the same time so ritualized that once one of the combatants is defeated, he is not pursued very far by the winner. Sexual rivalry occurs only during the breeding season, from December to March. At other times the bulls are sociable with one another.

G.C. Kelley

A pronghorn buck attempts to attract one of the does in his mating harem of a dozen. If she is not ready to reproduce, she will not respond to his advances. The females of very few species in the animal kingdom will accept a male unless they are biologically prepared to have their ova fertilized. Even if rejected by a doe, the pronghorn buck has plenty of things to do, because he must chase away rival males trying to steal his harem. Young males are easily driven off, but those similar in strength to the sultan of the harem must be fought. During the two days the photographer spent with this male and his does, the pronghorn buck battled and bested two different rivals.

Stephen Dalton

When the female broad-winged damselfly is ready to lay her eggs, the male clasps her head with appendages near his tail. Positioned vertically to her, he hovers in the air while she inserts her abdomen under aquatic vegetation and begins to deposit her eggs there. While the female is laying her eggs, the male rotates her in clockwise fashion so that the fertilized ova are deposited in a neat little circle. She may make several orbits, extending her abdomen further and further under the vegetation each time.

Robert T. Evans

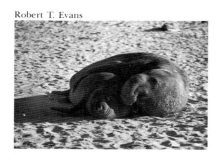

A female elephant seal is impregnated by a bull only three weeks after giving birth to the offspring she conceived the year before. Dominated by the male, she will be pregnant for most of her life. Even before the one season's baby elephant seals are weaned, she is biologically ready to conceive the young of the next year.

For many animals, the reproductive relationship between the sexes is a transitory thing. These flap-necked chameleons mated for about half an hour, then separated and went their own ways. The male has nothing more to do with the reproductive process, but the female must descend from her perch, dig a nest hole about eight inches deep and lay her clutch of almost three dozen eggs in it. They will hatch in three months.

Anthony Bannister

Two skippers mate. These insects, which generally are considered butterflies but also have mothlike traits, dart swiftly through the air. Some of the skippers engage in spectacular mating dances. Pairs zoom in arching flights of several yards at a time. A lively rhythmic-patterned succession of sexual movements are performed by many other types of butterflies. Typically, the male, intensely excited, approaches any likely-looking butterflies in his path until he finds a receptive female of his species. Then the preliminary flight begins, ending when she is sufficiently aroused. The two insects settle on a leaf or blade of grass and copulate, their wings opening and closing until the male fertilizes the eggs.

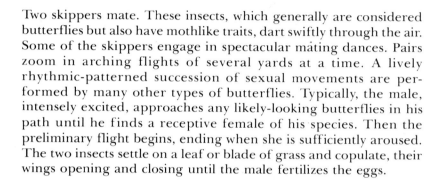

Zelda Glasser

Roman snails engage in foreplay prior to mating. Because both snails have male and female sexual organs, they will fertilize each other's ova. During copulation, each of these mollusks places its male sex organ in the female orifice of its mate, a maneuver that is possible only if the snails are approximately the same size.

K.P. Mafham

Francisco Erize

During mating, the dolphin gull utters a long, piercing scream. He will mount the female only after she has indicated her readiness by tossing her head and crouching in the mating position. The female will lay her eggs on the beach, where they, and later the young, are extremely vulnerable to predators. This vulnerability of the newborn—which are strong enough to run about shortly after being hatched—is offset by the nesting habit of their parents. Dolphin gulls, like most birds that raise their offspring on the beach, nest in large colonies. Myriad adults are on guard at all times, ready to sound the alarm if an enemy approaches.

A pair of lions will mate repeatedly, over a period of a few days. Lovemaking alternates with playful bouts of wrestling and nuzzling. The dominant male of a pride mates with all of the females in his group, each in her own time. He keeps the pride together and protects it, but the lionesses do virtually all the other work, helping the male hunt and rearing the many cubs that he sires. Although the females kill the prey, the male eats first, shoving his mates and young aside. The others get the food that remains, according to their size and strength. The male will rule until he is displaced, perhaps by one of his own sons that has grown and cannot be chased from the group, or more likely by a wandering bachelor. Struggles for supremacy often end in the death of one combatant.

2 New Generations

A marsupial tree frog rests on an agave leaf in Ecuador. In a striking departure from the amphibian norm, each female carries the fertilized eggs, and later the tadpoles that hatch from them, in a pouch on her back. In some species the tadpoles are freed from the pouch to complete their development in the water-filled cups of bromeliads, which are plants that perch on tree branches above the forest floor. In other species, when the eggs have large yolks, females transport their young until the tadpoles become froglets.

A pregnant cheetah surveys the landscape. This long-legged, swift-moving cat is normally a solitary animal, but when a female is pregnant, her mate may sometimes linger in her vicinity. The female gives birth after three months to a litter that usually numbers between two and four cubs, although in rare instances there may be as many as six offspring at a time. Cheetahs stay close to herds of gazelles, their favorite prey, so that when the female cat goes hunting, she generally does not have to stray too far from her young ones. Even if the mother is only a short distance away, however, her cubs are vulnerable to more aggressive predators, for—unlike a lioness—the cheetah does not belong to a pride, with baby-sitters always available to care for the young. The female cheetah herself must teach the offspring how to hunt. For example, she sometimes captures a young gazelle alive, carries it to her cubs and releases it for them to pursue.

Exhibiting the brood patch that appears when it is time to initiate the incubation process, a chin-strap penguin prepares to settle down on its egg. Unlike the mammalian embryo, which is sheltered within its mother's body, the developing bird in its egg is subject to the vagaries of weather, so that it must be kept warm by the proximity of the parent's body. For many bird species, the brood patch facilitates incubation. It is a swollen area of bare skin, richly supplied with blood vessels that generate considerable warmth. This patch comes into contact with the egg when the adult covers it. The distension of this small area of skin can become intensely uncomfortable for the parent, which seeks relief by sitting on the cool egg. In this way the patch both encourages an adult bird to begin the incubation process and provides the necessary warmth for the embryonic fledglings.

Francisco Erize

A white ibis nestling hatches, after first establishing contact with its parents by calling while still within the egg. Eggshell is one of the strongest materials in nature; its structure permits it to withstand outside pressure equivalent to ten tons per square inch. Yet this hard outer covering can be broken with ease by the weak hatchling when the time arrives for it to come out. Once it has emerged, the young ibis will be totally dependent upon its family for many weeks. The hatchlings of some other birds are considerably more developed, however, and are immediately ready to fend for themselves, although they still need parental protection and guidance until they grow larger.

Wendell D. Metzen

Encased by gelatinous eggs, embryonic spotted salamanders quickly develop into free-swimming larvae. These larvae hatch in the spring and spend the summer in the water, breathing by means of gills. Before winter the young salamanders have changed into adult form and have forsaken the water for the land. When they in turn are ready to breed, they will return to the water. The males will wheel about in a graceful reproductive dance and leave small, white sperm capsules on decaying leaves and twigs in shallow ponds and swamps. Then the females will swim to the bottom and take up the sperm capsules to fertilize the eggs. Mating accomplished, the adults return to the land, and the cycle begins again.

Edward R. Degginger

A miniature of the adult, the newly hatched hognose snake is ready for life on its own as soon as it emerges from the egg. If threatened, the little snake will dilate its neck, flatten its head and hiss—just as its parents do when they sense danger. If the threat does not cease in response to the young snake's aggressive appearance and action, the reptile will roll over on its back and play dead, exactly as a full-grown member of its species would do in a similar situation. The young are about five inches long upon hatching, and immediately disperse, feeding on insects until they are big enough to catch and eat toads, the main diet of the adults.

After hatching from eggs which have incubated in a mound of rank vegetation and mud, young alligators skitter for the water which always is nearby. The mother alligator often will help her offspring dig out of the nest. On hearing their cries in the mound, she generally will bite it open. Some mothers seem to stay fairly close to their young to protect them after hatching. However, any adult female or male alligator may respond to the shrill alarm cry of a newly hatched youngster and try to defend it.

While a day-old chick wanders about the rim of the nest, a parent saurus crane makes some adjustments in the structure of reeds and grass before continuing to brood. Parent cranes are strongly attached to one another and seldom are seen apart. Usually they travel with their young in small groups, which sometimes contain two or three families. During incubation, the female spends most of her time on the nest, which has been built on high ground in a marsh, while the male stands guard nearby. The strong, sharp beak of the adult saurus crane is an effective deterrent to small predators that might threaten eggs or young.

An aphid gives birth to live young that are tiny replicas of herself. The young, which are independent as soon as their feet touch plant stems, are all females, products of a form of virgin birth known as parthenogenesis. They result from eggs in which the embryos develop in the female without fertilization. Aphids give birth to live young all summer long. Toward autumn, some males are also born. They will mate with females, which then produce fertilized eggs before the cold weather kills off all members of this species. Although the insects die, the eggs survive the winter and, in the spring, give life to a new generation of aphids.

Still partly encased in the sack in which it was born, a young zebra is nuzzled by its mother. Immediately after birth, the mammalian mother begins to nose and lick her offspring. This behavior serves many biological purposes such as clearing the youngster's nose and mouth, drying its wet coat and triggering its first bowel movement. It also establishes intimate contact between the youngster and the parent, giving the newborn animal a sense of security, which from then on will be associated with the mother.

Norman Myers

3 Ties That Bind

Huddled beneath its towering parent, a greater flamingo chick rests after being fed. Its diet is a liquid secretion of algae, predigested by the parent and then drooled into the youngster's bill. Each feeding takes several minutes. Although flamingoes nest in colonies that may have several thousand families, the adults easily recognize their own young amid the throngs. Members of an individual family manage to stay together until the chicks can forage on their own.

M. Philip Kahl

Dwelling on the savanna, but never far from the edges of broken woodland, vervet monkeys live in a highly organized society in which the young are the objects of intense concern. Vervet mothers temporarily give over their youngsters to adolescent females to hold, so that these mothers of the future can learn the secrets of tender, loving handling. Later, when these adolescents mature and also give birth, they already will know how to carry their young ones gently. This apprenticeship in maternal behavior—a practice that is typical in monkey society—will benefit both new babies and new mothers. This process of learning by doing and learning by imitating is carefully supervised by the older female, which never allows an adolescent to touch her offspring until she knows that the baby is strong enough not to be hurt seriously if accidentally jounced or dropped.

Arthur Christiansen

Edward S. Ross

A lioness nurses her cubs. If necessary, she will nourish and nurture the offspring of other females in the pride, for lionesses cooperate in rearing the group's young. This common effort for the common good increases the chances of the cubs surviving, for it provides greater protection against enemies. Hyenas, leopards and even wandering lions that do not belong to the pride will snatch up a cub if given the chance. Despite the care it receives, each young lion is lucky if it stays alive during its first year. Most die, especially if food is scarce, because the youngest and weakest in the pride are the last in line to eat. Starvation is the greatest threat to a lion throughout its existence. Well-fed captive lions live as long as thirty years, but in the wild only those that are exceptional survive for even half that length of time.

James Malcolm

African hunting dogs, which live together with a remarkable degree of cooperativeness, show unusual sensitivity to the needs of the young in their pack. Females vie with one another for the privilege of nursing the pups, although mothers retain first rights. Even the males play an important part in rearing youngsters, which average about ten to a litter. In one observed case, a litter of nine pups was orphaned when five weeks of age. The little ones were raised by the other members of the pack, which—after the death of the pups' mother—were all males. Until the young are strong enough to follow the grown-ups on a hunt, the adults regurgitate food for them at the den.

Leonard Zorn

A parent green heron has arrived at the nest with food, and the young fledglings hustle out, fighting to be first to eat. The young do not all hatch simultaneously, so some are larger than others, a distinct advantage when it comes to shoving away competitors at mealtime. The youngsters are so voracious that feeding them is a full-time task for the parents, who must continually search for food, even though the breeding season coincides with the time when prey is most numerous. Should drought or some other disaster eliminate the small fish and other delicacies that herons favor, this search for sustenance would still proceed, for the members of this species cannot think and cannot modify their own behavior. The parents only know that they must go on and on and on, seeking food to still the cries of their hungry young.

Arthur Christiansen

When other adult ducks were frightened and flew away, this female shelduck gathered several broods of ducklings and led them to safety in the reeds of a brackish Danish lagoon. If this had occurred some weeks earlier, only the ducklings in the female's own brood might have followed her. Newly hatched ducklings identify so strongly with their mother that she is the only creature, other than one another, with which they will associate. This is brought about through a process called "imprinting," which begins even while the ducklings are in the eggs. Then, the first thing they hear is their mother. On hatching, she is the first large, moving object they see and they instinctively follow her. Later, this behavior extends to others of their own species, helping them to associate with their own kind for activities such as feeding and breeding.

Born in a den dug by its mother in a snowbank, a young polar bear is at first helpless and sightless. Its eyes open at about a month, but it still needs its mother's care for a long period after that. It is estimated that the cubs remain with their mother for at least a year, and in some cases twice as long as that time. At birth, a youngster is clothed in fine white hair, weighs about a pound and is less than a foot long. Mature polar bears can reach more than sixteen hundred pounds and stand as high as ten feet, although most are somewhat smaller.

Fred Bruemmer

Content and secure, a young hippopotamus is tended by its huge mother, while others of the herd laze in the waters nearby. Highly social, hippos live in groups of up to more than a dozen animals, with females and youngsters at the center and the males at the periphery. Except when the population of this species becomes unusually dense, the adults seldom act aggressively toward one another. Under the stress of a population explosion, however, males often fight each other. Otherwise, life for the young hippo is rather quiet. The hippo is born underwater and must be lifted to the surface by its mother so it can breathe, but it soon learns to float and swim without help.

Norman Myers

An Alaskan brown bear helps her cub cope with the rough waters of the McNeil River. It is early summer and a number of bears, normally solitary, have congregated for the annual salmon run. The huge, shambling creatures, largest living carnivores, compete for the best fishing spots and the most salmon. The cubs tag along behind their mothers, but wait ashore or in the shallows when the females fish. Often the youngsters play in groups. Very young cubs seem to get anxious if their mothers leave them for too long and will attach themselves to any female that happens by. The foster-mother often accepts the foundling, and will even nurse it; but if the real mother returns, the cub will be reclaimed, though not always without a fight.

Jeff Foott

Richard Wrangham

Totally helpless at birth, the young orangutan will not be weaned until it is more than three years old. Before that time, however, the mother begins to feed her baby solid food that she has first chewed to a pulp. As the infant gains strength, the mother guides it in its first movements about the branches of the trees in which they live. Orangutans are the only great apes that are truly arboreal. To survive, they must have deep forest with tall, mature trees. Destruction of such dense woodlands in the animal's range on the islands of Borneo and Sumatra has placed the species in extreme danger. Orangutans also have been reduced in numbers because, until recently, the standard way for a human being to capture one was to shoot a mother and take her infant.

Joseph Popp

An adult vervet monkey puts an arm around a young one. Perhaps only among primates is such a gesture fraught with so much meaning. The young vervet is only in the process of being weaned and, like other primates, has a long stretch of time ahead of it in which to grow up. At the present stage of its development, the little one benefits from the affection of its mother and other grown-ups in the group. These contacts help the vervet adjust to living with the others in the troop and teach it to make adult attachments—such as sexual and parental bonds—later on. All these relationships go toward building the troop into a unit that promotes the survival, not only of its members, but also of the entire species.

Al Szabo

Surrounded by solicitous females, an elephant calf is being taught to drink at the water hole. The calf's trunk is too short to reach the water from a standing position, so the older elephants will gently nudge the youngster into a kneeling posture. Any one of the females in the herd may act as a surrogate parent, and young elephants are free to suckle from any nursing mother. Adolescent females, who have not yet reached sexual maturity, learn the kind of maternal behavior that will be required of them later as they baby-sit for playful youngsters.

4 Play and Perils

James K. Morgan

Hiding amid the flowers of its summer grazing grounds, a bighorn lamb rests for a moment from play. The lamb spends its first summer frolicking with other newborn in the herd. The young wild sheep prances with ease over mountain meadows and craggy outcrops, for it instinctively knows how to move confidently about in high places without falling. While the lambs gambol and romp, their mothers keep a watchful eye on them. All the ewes share in tending the young, but mothers will suckle only their own offspring.

A newborn harp seal is snow-white, weighs about a dozen pounds and cannot swim. In two weeks the seal will have changed drastically. It then weighs as much as eighty pounds, its coat has darkened and it is voraciously hungry because its mother has deserted it to join males that are disporting themselves in the water just off the breeding grounds. Hunger eventually drives the young seal into the water, in which, with its new strength and size, the youngster can learn to maneuver expertly. At first it swims poorly, only well enough to catch slow-moving crustaceans. Soon it will be able to pursue and catch fish.

Fred Bruemmer

On an ice-covered tundra lake, a caribou calf takes a rare moment of rest during its vigorous and hazardous first few months of life. Its mother and the rest of the herd are nearby—a necessity if the calf is to have any chance of surviving. Wolf packs follow the herds, waiting to pick off isolated stragglers. In summer, bloodsucking insects that plague the adults can kill a young caribou unless its mother is there to wipe them from its face. The calf has just one short summer to develop the strength it needs to withstand the Arctic winter.

Fred Bruemmer

Born blind and helpless in a den, bobcats are weaned and ready to forage with their mother by the time they are two months old. If caught in the open away from parental protection, the kits can be seized and devoured by great horned owls, dogs and other predators. For their first summer, the bobcat kittens stay with their mother, learning to stalk and capture increasingly larger animals. By early winter they will be ready to go out on their own in search of hunting grounds.

Leonard Lee Rue III

After red fox pups are little more than a month old, they emerge from the den and play before it for hours at a time. Later, they will move much of their play to another site some distance from the den, called by scientists a "rallying station." After about ten weeks, the young are weaned and live on food brought back by the parents from their foraging. These pups are the progeny of a mother whose coloration is typical of her sex—and a father with unusual color for the species. Because his markings include a dark cross on his back and shoulders, he is given the name "cross fox," which in no way refers to his disposition.

Helen Rhode

Great white heron nestlings freeze in response to an alarm call from a nearby parent. Their reaction is not reasoned, but instinctive, and is characteristic of many young birds. It helps the youngsters escape detection by predators. Later in life, this ability to keep as still as statues will assist the herons as they stalk fish and other small aquatic creatures in the shallows.

Laura Riley

Leonard Lee Rue III

A black bear cub wanders alone, but only for the moment, because its mother is nearby. For two years the young bear will learn the basics of survival as it travels with its mother around her territory, several miles in area. The cub will use its sensitive nose to sniff out acorns, berries and grubs to eat. It will be with the mother as she selects a snug den for the winter—not to hibernate, but to sleep for long periods when the weather is especially severe. At the end of the second year, the youngster will leave its mother. If a male, its time as a member of a family will have ended, for except during the mating period, male bears are solitary wanderers. If a female, the cub will have one solitary summer before she mates and has youngsters of her own to care for; overall, her fertility period will last for ten years.

Fran Hall

Six-feet tall and more than 150 pounds at birth, the young giraffe can run on its own before it is a day old, but it needs to rest frequently. The youngster grows up in a herd, which includes adults of both sexes. Females cooperate with one another in watching over the calves and share with the bulls the responsibility of protecting the young from predators. Although adults are vulnerable to interlopers only when lying down and asleep, young giraffes are easy prey. When a calf is threatened, the adults defend it with pile-driver kicks, which can kill even a lion.

David Agee

At about two months of age, African hunting dog pups can accompany their pack in pursuit of game. The adults show such concern for the youngsters that the pups are given precedence at a kill. By following the pack during a hunt, the young dogs learn the highly complex hunting tactics of their elders, including decoying and ambushing.

Dean Krakel II

Learning to butt is part of growing up for a five-month-old male elk. He plays at the behavior that later in life he will use in combats with others of his own sex during competition for the right to mate with cows. Although violent, these bouts generally amount to no more than heavy pushing and shoving, with antlers often entangling or interlocking, rather than fencing for a dangerous stab to the body.

The first social contacts made by a young baboon are with its mother's older offspring. Within six weeks after birth, the baboon spends most of each day with a small, active band of other infants and juveniles. A social hierarchy develops among the youngsters that will change hardly at all as they grow into adults. The social links that form in these childhood groups will, in many cases, last for life. For these young baboons, juvenile play is a method of testing their strength and its limits and a way of learning how to control aggression, a requisite for cooperative existence in the troop. Experiments have shown that macaque monkeys—similar to baboons—grow up to be socially unbalanced adults if reared away from other young of their kind.

Al Szabo

A trio of lion cubs sets out for a short journey at the edges of the pride. During its two years of childhood, each cub spends much time socializing with its littermates and other young lions. The time is well spent. Together, through play, the youngsters build the bonds that will help them become full-fledged members of the pride and learn from experience the techniques they will need to survive as adults. The female cubs gradually find out how to cooperate in rearing the young and hunting for prey. The playing of the male cubs represents a kind of rehearsal for the actual struggles for dominance that will occur when, as adults, they strive for the right to head a pride of their own.

Fran Hall

A feisty, furry ball of activity and curiosity, the young raccoon seems constantly on the move, except when sleeping among its littermates in a heap within the snug confines of its den. During its first summer, it will accompany its mother on hunting forays, learning to search for and find food. By the end of summer, the bonds between mother and offspring are severed, and by winter the family has broken up. Then the youngsters must use what they have learned—and what their instincts tell them—to fend for themselves.

Maggie Cavagnaro

5 Checks and Balances

Robert T. Evans

These two young northern elephant seals belong to a species that is unusually sociable. Vast number of these animals flop together on rocky beaches, seldom fighting except when bulls compete in the breeding season. However, a considerable amount of jostling goes on as the seals nestle one against the other. The elephant seals are so companionable that they often snuggle up to California sea lions, which use the same beaches on islands off Southern California and Baja.

Steven C. Wilson

Bob Campbell

Francisco Erize

With wildebeest in the background, a large zebra herd—an extended family group that may number more than a dozen individuals—gathers around a water hole. The herd, consisting of mares and young, is headed by a stallion, although it is not dependent upon his presence for cohesion. If the head of the herd dies, the others will stay together under the leadership of a dominant female until they are eventually taken over by another stallion. Young males often steal mares from established groups, so a leaderless zebra aggregation may lose a few members before it comes under new rule. Zebras mix with wildebeest, eland and many other hoofed animals on the African savanna. Conflict is at a minimum. The herds of the different species are independent of one another but stay together; perhaps there is safety in numbers—the more eyes and ears on the lookout the better.

A sight that was common in the past but is rare today is that of American bison and pronghorn antelope grazing on the prairie. Both creatures were slaughtered by the millions in the United States in the last half of the nineteenth century. Herbage-eating animals often tolerate other grazers in their neighborhood, and many readily mix. The pronghorn and bison both benefit from their association: The buffalo is larger and stronger, whereas the antelope is much more alert. This combination of power and watchfulness is an effective deterrent to any predators. When a pronghorn sees an enemy, it erects the long hair of its white rump. This patch of hair flashes in the sun and can be observed across the open prairie from much more than a mile away. While the antelope is sending out this warning, it is also releasing a strong scent from glands in this same rump patch—an odor so potent that it can spread for hundreds of yards. This double signal will send other pronghorns as well as bison into a headlong rush for safety.

King penguins gather by the thousands to nest on a subpolar island. They are able to live in such numbers because the colony is governed by instinctive rules. Although the birds seek out the company of their own kind, each pair requires its separate nesting territory, no matter how small. The area is about a yard square, a spacing that is rigidly maintained. If the territory is violated by a neighbor, a squabble erupts, but it ends as soon as the interloper retreats beyond the boundary, keeping conflict at a minimum among the penguin throngs. King penguins do not build nests but, instead, place their egg on the ground. Male and female share incubation, during which the egg is shielded from the cold by a flap of skin on the belly of the parent. The large, fuzzy chicks depend on their parents for almost a year, and feeding the babies takes the combined fishing ability of both mother and father. When the parents go off to fetch food, the territorial restrictions relax, enabling the youngsters to cluster together under the eye of adults who have stayed behind to baby-sit. Returning penguins are able to find their own offspring among myriad look-alikes by calling, which in some unknown manner brings the young to the right nest.

Hamadryas baboons gather for the night. These apes, which inhabit mountainous terrain in Ethiopia, do not spend the day in large troops but instead forage in small family units, which have only one adult male. As darkness approaches, the baboon families congregate, sometimes by the hundreds, and sleep high on the rocks. By splitting up during the daytime, the baboons decrease the chances of their fighting over the meager food available in their stony upland home. By gathering at night, they increase their ability to ward off predators.

Joseph Popp

Pacific walruses bask on an Alaskan beach. They are highly social creatures, with a strong sense of herd togetherness. A hunter or polar bear stalking a single walrus runs the risk of being savagely attacked by the entire group. The calf of this species remains with its mother for two years, the time it takes for the youngster to develop the strong pre-molars it needs to crush the clams and other shellfish that provide much of its diet. Pacific walruses follow the seasonal buildup and diminishment of the Arctic ice pack, often riding along on swift-moving floes. Sometimes an entire herd will leave a sheet of floating ice and haul up on a beach to sunbathe. When a walrus is out of the water, its body temperature rises. In compensation, the blood vessels near the surface of the creature's body start to dilate, allowing excess heat to escape. This causes the skin to turn pink, a color easily visible in older walruses who have lost most of their body hair.

Fred Bruemmer

A relative of the European chamois, but even more adept at living on steep, rock elevations, the Rocky Mountain goat has hooves with hard, thin edges and treaded soles, natural nonskid equipment. The mountain goats—misnamed because actually they are not of the goat species—breed in the autumn. Offspring, which weigh about six pounds each, are born the following spring. During that season and in the summer, the females and youngsters herd together, moving without an apparent leader over the highland feeding grounds. The males form herds of their own until the time to mate. The bachelor groups generally wander to the steepest parts of the mountains, while the females and the young stay slightly lower down, where food is more ample. A goat youngster develops quickly, so that by the middle of the summer it weighs about forty pounds and can go anywhere with its mother.

Steven C. Wilson

The wolf pack generally contains less than ten animals but may have as many as fifty (although this is rare). Large packs often break up into smaller bands, especially during the summer. Where wolves are scarce, as they are in parts of southern Europe, sometimes as few as two or three animals band together. Packs of about a dozen seem to have a better chance at bringing down larger prey, such as red deer and moose. In hunting and killing animals for food, wolves employ cooperative tactics. If a pack has too many members, however, a single kill, even of a thousand-pound moose, may not provide enough meat to go around.

Rolf O. Peterson

Sonja Bullaty

Edward S. Ross

Leonard Lee Rue III

Joseph Popp

Fleeing before the approaching cold, snow geese arrive at a rest stop on their way from the northern tundra to southern wintering havens. Only a short time before, these birds were all flightless. The young had not yet grown their flight feathers, whereas the adults, temporarily grounded, were losing theirs—shedding old, worn feathers and replacing them with new. Although most birds molt a little at a time, without losing the power of flight, geese are among the minority that change all their feathers at once. Even when flightless, snow geese still can cope with predators because of their numbers, aggressive behavior and powerful bills.

Giraffes on the move can cover even rough ground at thirty-five miles an hour. Herds often number more than seventy animals, but these large associations are subdivided into family groups, headed by a dominant bull, with subordinate males, females and youngsters. Because giraffes possess good eyesight and great height, they can keep in contact with one another even when separated by wide stretches of savanna, so sometimes their groups and herds become widely dispersed. Zebras and larger antelopes that occasionally accompany giraffes seem to depend on the tall creatures as sentinels.

Agitated by the approach of a photographer, cow elephants gather instinctively about their young to face the assumed danger. The largest elephant is the herd matriarch, the leader and chief protector. If the group must retreat, she serves as a rear guard, always staying behind the slowest calves. She uses bluff, and if that fails, an actual charge to fend off the enemy. Small family groups sometimes loosely link up with larger herds, made up of females, calves and immature males. The bulls occasionally join the herds but more often stay around their fringes, attaching themselves to individual cows when they are ready to mate. As a rule, even before a matriarch dies, her successor gradually assumes leadership, but if the matriarch were to be killed unexpectedly, the cohesion of the group might break down.

6 The Closed Circle

A male Hamadryas baboon is distressed at finding a dead young baboon. In this species, infants evoke a deep-seated maternal and paternal response in adults. When the troop is on the move, the big males escort the mothers with small young and are ready to defend them against predators. In turn, the mothers seek out the protection of the dominant males. All such behavior helps baboon babies during their long infancy, which, in primates, typically lasts at least one year.

A never-ending cyclical relationship exists between predators and prey. The young buffalo that has been killed by these lions has lost its life so that the big cats may survive. If predator populations are dense, flesh-eaters will regularly thin out the weak, sick and otherwise unfit from herds of game, keeping the prey species from overpopulating and starving. A year in which the game herds diminish drastically is usually followed by one in which the number of predators declines sharply. This drop in the predator population in turn permits the game herds to rebuild. The relationship between prey and predators helps to keep the natural world in balance, and the cycles operate smoothly, except when humans interfere by killing off predators and causing game to become too numerous for its own good.

Irven Devore

A herd of American bison, facing into the blowing snow, wait out a winter storm in Yellowstone Natural Park. When a violent disturbance of the atmosphere threatens, these large animals may at first attempt to avoid it by moving before the weather, or they may try to find a stand of trees or some other haven from the wind. If caught by surprise in the open, however, they are able to rough it out, because during the autumn the buffalo undergoes physiological changes that promote survival in the harsh winter of the North American plains. Gorged on the lush grasses of late summer, the bison adds a layer of insulating fat to its body, which grows a heavy, shaggy winter coat.

Steven C. Wilson

Gregarious and vocal, sea lions gather in large colonies along the Pacific coast of North America. They feed voraciously on fish and squid, taking some commercial species but also many not used by man. While it is true that sea lions occasionally devour salmon, these marine mammals greedily eat lampreys as well, which are the bane of salmon. Effective predators, sea lions themselves fall prey to the most savage natural hunters in the sea—sharks and killer whales.

Steven C. Wilson

Pelted by freezing rain, a lone, weakened caribou sinks to the tundra in the Brooks Range of Alaska, never to rise again. Despite such individual disasters, this wild creature is so well adapted to sub-zero weather that the species has changed little since the ice ages of the late Pleistocene era. The metabolism of the caribou—the same species as the Eurasian reindeer—is such that it can withstand temperatures as extreme as ninety degrees below zero and can thrive on a diet of lichens so low in protein that it would be fatal to other hoofed mammals.

Steven C. Wilson

James K. Morgan

Silhouetted against the moon, a Rocky Mountain bighorn ram is the culmination of perhaps ten million years of evolution, which has adapted this splendid creature for life on the high places of western North America. Even so, existence for the ram is a continual struggle. In summer, he must find adequate nourishment among the crags; in winter, when the snows drive him from the heights, he must seek food down in the lowland valleys. He must strive to mate, slamming his magnificent horns against those of other males in ritual combats. It is for these highly structured battles that his horns have evolved. A bighorn is a serious contender in mating contests for only a few of his fourteen years. Only the fittest males, creatures in their prime with great horns, win the females, assuring that their superior traits will be passed on genetically to the next generation. The evolutionary lottery, which has made the dominant males perfect specimens of their kind, works extremely well under purely natural conditions. But ironically, when human beings enter the picture the magnificent horns that signify a ram's dominance are those which attract trophy hunters, and therefore the best of the breed are often destroyed before their full genetic potential is reached.

Steven C. Wilson

A redwing blackbird has returned to North Dakota after wintering in the south and is now tending its nest and eggs. In the northern part of the United States, the redwing is one of the first heralds of the coming spring, often arriving in the final days of winter when snow still covers the ground. Even on a chilling morning, the loud trill of this bird is a reminder that, although the world looks glacial, the seasonal cycle is moving from winter's dormancy and cold to spring's first flush of renewed life.

Steven C. Wilson

An American bison, alone on the bleak plains, symbolizes the tragic history of its kind. Far back in the Pleistocene era, its ancestors crossed the land bridge linking Asia and Alaska and arrived at grasslands that provided a virtually endless food supply. The numbers of the buffalo swelled to millions upon millions. Prehistoric peoples followed the great herds; later, entire Indian cultures centered upon them. Later still, the meat of the bison fed the first whites who ventured into the forests on the western side of the Appalachians and then pushed on to the prairies and plains that stretched toward the sunset. When the westward movement accelerated, the buffalo was doomed. The herds were slaughtered for their hides, for fun, for relief from boredom, for profit. And once the bison was exterminated, the Plains Indians either starved or were forced to accept reservation beef. In the eyes of the advancing whites, there was no room for either the Indians who lived in harmony with the land rather than trying to tame it, or for the vast herds of huge, stolid animals that roamed at will and could not respect fences or differentiate between grass and wheat. It took only a few decades to destroy more than sixty million bison. Today, remnant herds survive in parks and preserves, managed much like domestic cattle. Today, the remaining Indians are mostly confined to reservations. It is not the same.

Cormorants flock in Japan. Creatures of coastal waters, lakes and rivers, these dark-colored, web-footed seabirds have existed for at least fifty million years. They are fishers, chasing slower fish underwater to depths of about a hundred feet. Cormorants have been known to live for almost twenty years, and it is believed that some may survive to even twice that age. They gather in colonies of thousands to nest on rocks near the water. The future of their kind depends on these rookeries, which grow increasingly scarce as the wild, unpolluted seacoasts of the world disappear.

Tojo Tanaka

DECORATIVE PLATES

1   A white-footed mouse (*Peromyscus leucopus*) peeks out from its down-filled nest. United States. Robert Carr.

2   A Thomson's gazelle (*Gazella thomsoni*) silhouetted against the sunrise. Africa. M. Philip Kahl.

5   Vervet monkeys (*Ceropithecus aethiops*) in the rain. Africa. Edward S. Ross.

6   A male Caspian tern (*Sterna caspia*) presents its mate with a gift of fish. United States. Jeff Foott.

8   A skipper butterfly (*Hesperiidae*). United States. Edward R. Degginger.

10   A long-horned milkweed beetle (*Tetraopes tetraophalmus*). United States. Robert Carr.

15   A male lion (*Panthero leo*). Africa. Thase Daniel.

16   Rocky Mountain bighorn rams (*Ovis canadensis*) butt horns. United States. Marty Stouffer.

36   A hermit thrush (*Catharus guttatus*) nest. United States. Larry West.

56   A cheetah (*Acinonyx jubatus*) cub rests its head on its mother's back. Africa. Animals, Animals—Fran Hall.

76   A bighorn lamb (*Ovis canadensis*) with its mother. United States. James K. Morgan.

90   Two hippos (*Hippopotamus amphibius*) leave a water hole. Africa. Edward S. Ross.

108   A black-headed gull (*Larus ridibundus*). United States. Laura Riley.

130   A Galápagos land tortoise (*Testudo elephantopus*). Galápagos Islands. Tui De Roy Moore.

152   Giraffe (*giraffa camelopardalis*). Africa. Sonja Bullaty.

an AG Edition